Preface:

This book is a condensed explanation of power supplies used in the electronic communication industry. Chapter 1 is dedicated to linear power supply design. It includes half wave and full wave rectification, power supply ripple, regulated and unregulated supplies, capacitor multipliers, current limiting, and bootstrapping. Chapter 2 introduces switching power supply, both in theory and with actual circuits for buck, boost, flyback designs, and FCC emission requirements. Chapter 3 encompasses special power supply configurations such as diode voltage doublers and bandgap reference theory. Chapter 4 includes power generation techniques such as solar power, and wind turbines.

E

Introduction of Linear and Switching Power Supplies

Definitions

Transformer: Transforms 110 volts alternating current (AC) to a lower voltage AC, and provides isolation from the 110 VAC plug to the circuit, for safety.

Diode: Allows the electric current to flow only in one direction, thereby converting AC to Direct Current (DC)

Capacitor: Stores energy so that when it is not being charged on the negative AC half cycle, it can provide power to the circuit (Rload).

Rload: This is a resistance to simulate the resistance of the circuit that is powered by the power supply.

Linear Positive Supply

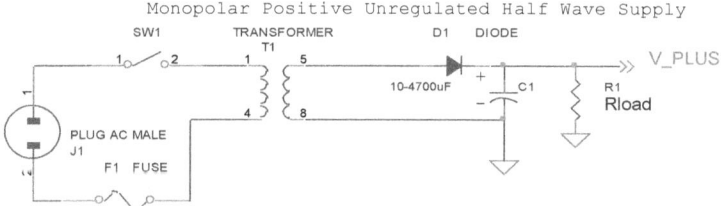

Figure 1: Basic Positive Linear Unregulated Half Wave Power Supply

Figure 2: Top row from left to right 1) power cord, 2) toggle switch, 3) fuse holder, 4) fuse: Bottom row from left to right, 5) Power transformer, 6) diode, 7) diode bridge, 8) Capacitor

Linear Negative Supply

In Figure 3, the diode is turned around and the capacitor is reversed to provide a negative dc voltage output.

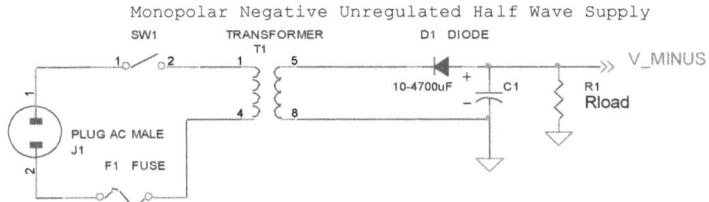

Figure 3: Monopolar Linear Negative Unregulated Half Wave Power Supply

Full Wave Supply with no Center Tap

Figure 4 shows how to use the same transformer to provide both positive and negative voltage without sacrificing current drive capability for either voltage.

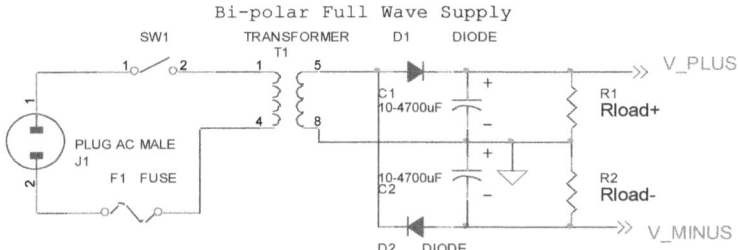

Figure 4: Bi-Polar Linear Full Wave Power Supply with unregulated positive and negative outputs.

Full Wave Supply with Bridge Diode and no Center Tap

In Figure 5, a full wave bridge utilizes both the positive half cycle and the negative half cycle to produce a positive output voltage.

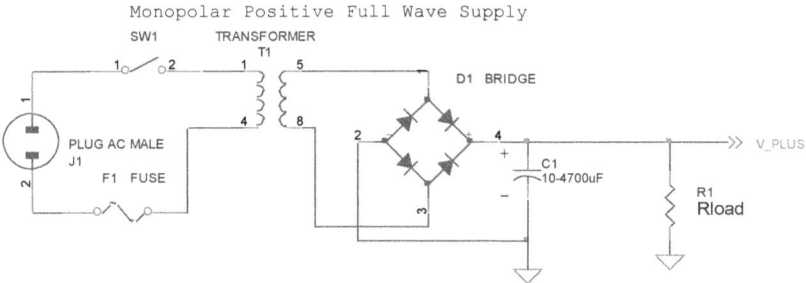

Figure 5: Monopolar Linear Positive Full Wave Unregulated Power Supply

Full Wave Supply with Bridge Diode and Center Tap

Figure 6 uses a full wave bridge with a center tap on the secondary of the transformer to produce both positive and negative voltage outputs.

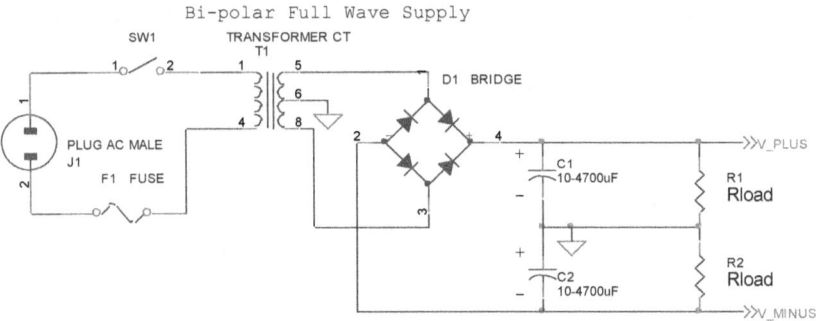

Figure 6: Bi-Polar Linear Full Wave Unregulated Power Supply

Power Supply Ripple

In Figure 7, the red curve shows the variation (ripple) in the dc output voltage. If the capacitor is removed, then the output voltage is represented by the dotted line. The capacitor is an energy storage device that acts to smooth the output voltage ripple that powers Rload (the electronic circuit).

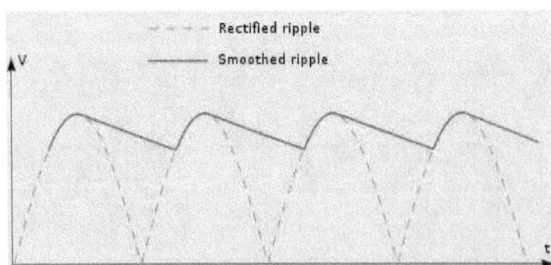

Figure 7: Ripple on the unregulated Linear power supply at the Load resistance

For a full-wave rectifier:

$$V_{pp} = \frac{1}{2fC}$$

For a half-wave rectification:

$$V_{pp} = \frac{1}{fc}$$

such that

- Vpp is the peak-to-peak ripple voltage
- I is the current in the circuit through the load resistance
- F is the frequency of the ac power
- C is the capacitance

Regulated Linear Power Supply

In Figure 8, a simple regulator is added to eliminate the ripple. Since a 4.3 volt zener diode was chosen, and the base emitter drop of Q1 is about 0.7 volts, then the output voltage will be 5.0 volts. This regulator has no feedback mechanism so the regulation at 5.0 volts is poor.

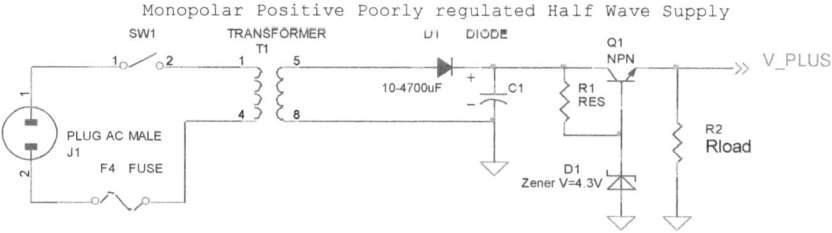

Figure 8: Simple Linear regulated power supply

Regulated Linear Power Supply with Feedback

By adding feedback, in Figure 9, the output voltage can be held tightly to a single voltage, by potentiometer R2, regardless of the value of Rload.

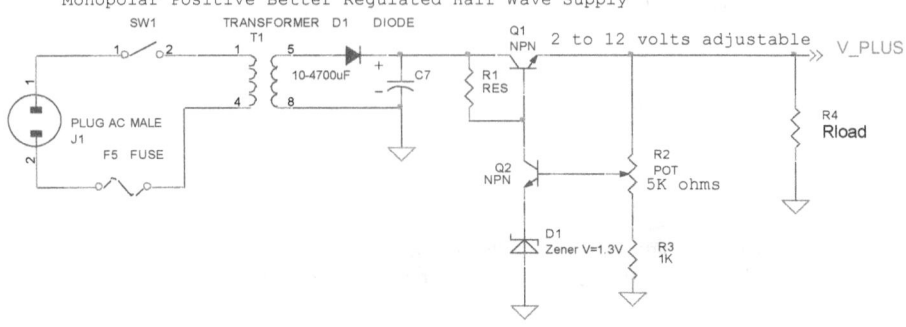

Figure 9: Simple Linear regulated linear power supply with feedback regulation.

Regulated Linear Power Supply with Feedback and Capacitance Multiplier

A current multiplier is shown in Figure 10. The ripple voltage is smoothed by the action of R1 and C1. This smoothed voltage is applied to the base of Q1. The voltage at the emitter of Q1 must echo the voltage at the base, minus a 0.7 volt voltage drop. Assuming R1 = 100 ohms and C1 = 1000μF and a ripple frequency of 60 Hz, the reduction of ripple is given by the voltage division ratio

$$VC1 = \frac{1}{1 + jR1 * 2 * \pi * 60 * C1}$$ where $j = \sqrt{-1}$ indicating that the ac current through

C1 leads the voltage across C1 by 90 degrees. Therefore the reduction of ripple is 2π*60*100*0.001 = 37.7.

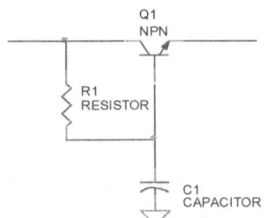

Figure 10: Capacitance Multiplier

In Figure 11, the current multiplier is added to the power supply. In reality, current multipliers are not used in modern power supplies.

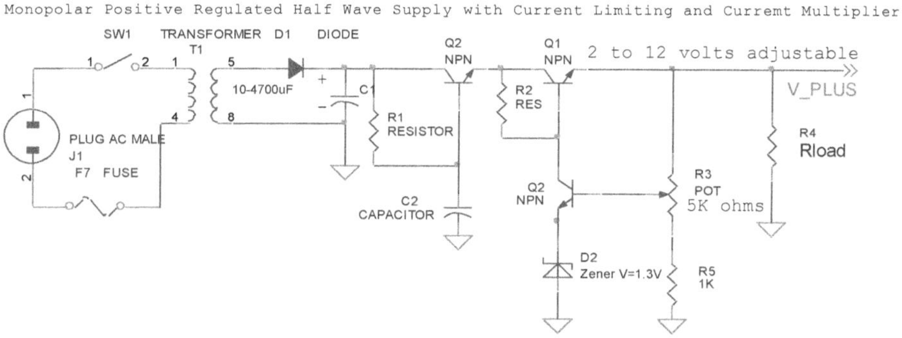

Figure 11: Linear Power supply with added current multiplier.

Regulated Linear Power Supply with Feedback and Capacitance Multiplier and Current Limiting

Figure 12 adds current limiting to the power supply. If the current exceeds 1 amp, then the voltage across R5 will be 0.6 volts, turning on Q3, which, through the action of Q4, lowers the base drive to Q1. Therefore, if the current tries to exceed 1 amp, the output voltage is lowered in order to maintain a 1 amp output current to Rload. A power supply operating in this mode is called a constant current power supply.

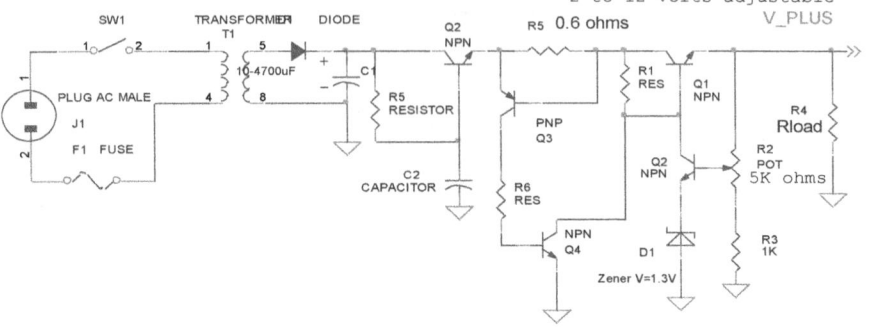

Figure 12: Simple linear positive half wave power supply includes current multiplier.

Linear Power Supply with Adjustment to 0 volts.

In previous figures, the output voltage was necessarily at least 0.6 volts lower than the voltage applied to the collector of Q1. By reversing the emitter and collector of Q1 (and other necessary changes) the maximum output voltage can be a few tenths of a volt higher because the collector-emitter saturation voltage of Q1 is approximately 0.2 volts. In practice, a metal oxide semiconductor field effect transistor (MOSFET) is normally used for Q1 because its drain to source on-resistance can be low enough so that the MOSFET drops less than a tenth of a volt. Figure 13 shows the reversal of the emitter-collector to create an LDO.

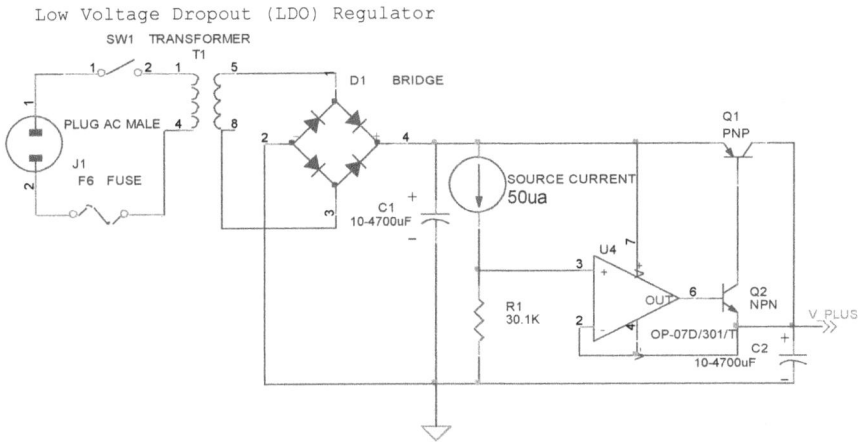

Low Voltage Dropout (LDO) Regulator

Figure 13: Simple linear positive full wave low voltage dropout (LDO) linear power supply adaptation for 1.5 volt output

Working Circuit of Monopolar Linear Power Supply

With reference to the schematic in Figure 14, lamp, LP2, is a power-on indicator. The other lamp (lower) lights when the unit reaches its preset current limit. R5, C2, and Q10 (TO-3 case) operate as a capacitor multiplier. The 36 volt zener across C2 limits the maximum supply voltage to the op-amps supply pins. D5, C4, C5, R15, and R16 provide a small amount of negative supply for the op-amps so that the op-amps can operate down to zero volts at the output pins (pins 6). A more modern design might eliminate these 4 components and use a CMOS rail-to-rail op-amp. Current limit is set by R3, D1, R4, R6, Q12, R10, and R13 providing a bias to U2 that partially turns off transistors Q9 and Q11 when the current limit is reached. R4 is a front panel potentiometer that sets the current limit, R22 is a front panel potentiometer that sets the output voltage (0-30 volts), and R11 is an internal trim-pot for calibration. The meter is a 1 milliamp meter with an internal resistance of 40 ohms. Switch S1 determines whether the meter reads 0-30 volts, or 0-1 amp.

9

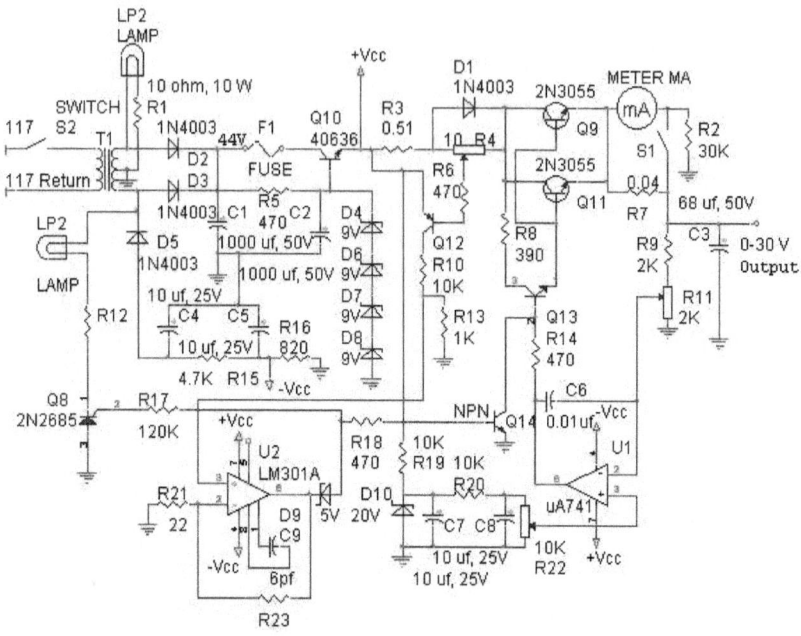

Figure 14: A complete monopolar linear regulated power supply

Working Circuit of Bi-polar Linear Power Supply

With reference to the schematic in Figure 15, lamp, LP2, is a power-on indicator. There is no adjustable current limiter in this unit, although R1, D2, and D3 set the current limit to approximately 0.25 amps on the positive side, and R15, D9 and D10 set the current limit to slightly greater than 0.25 amps on the negative side. The voltages at the output of Q1, and Q3 are pre-regulated to 20 volts so that the op-amp supply voltages do not exceed the maximum supply ratings for these op-amps. R9, R12 and U2 cause the negative supply to track the positive output voltage. R7 is a front panel potentiometer to adjust the output voltage(s).

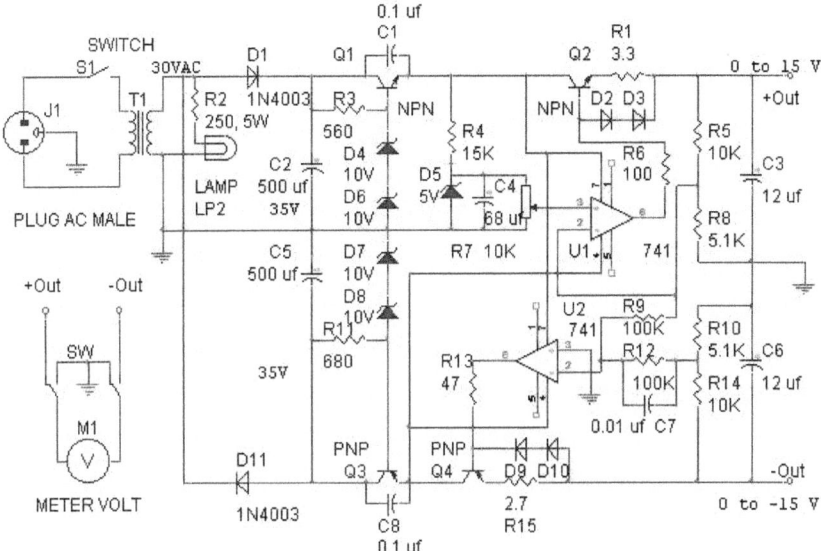

Figure 15: A complete bi-polar linear power supply.

Generic Model of Linear Power Supply

In practice, no one today builds power supplies from discrete transistor circuit. Figure 16 show a 3-terminal integrated circuit (IC) linear voltage regulator that performs all the functions of the discrete transistor circuit, but includes automatic shutoff when the integrated circuit temperature becomes too high, and automatic current foldback if the current tries to exceed the maximum rated current for the IC linear voltage regulator.

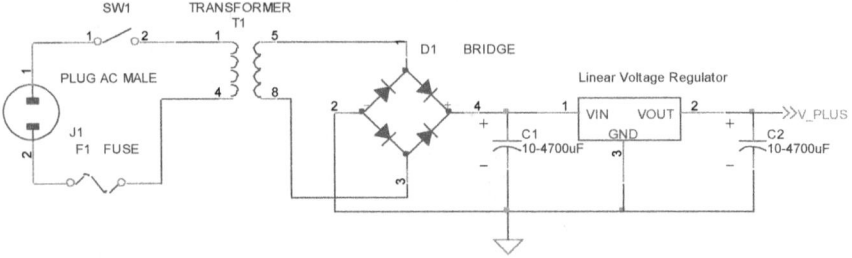

Figure 16: Generic Linear Power Supply Model

Bootstrapping a Linear Power Supply

If the current capability of the linear voltage regulator IC is insufficient, then it can be "bootstrapped" as shown in Figure 17. In this example, when the voltage across R1 exceeds 0.6 volts (1 amp current), then Q2 turns on to allow any current over 1 amp to flow through Q2 to the output. In practice, bootstrapping is seldom used.

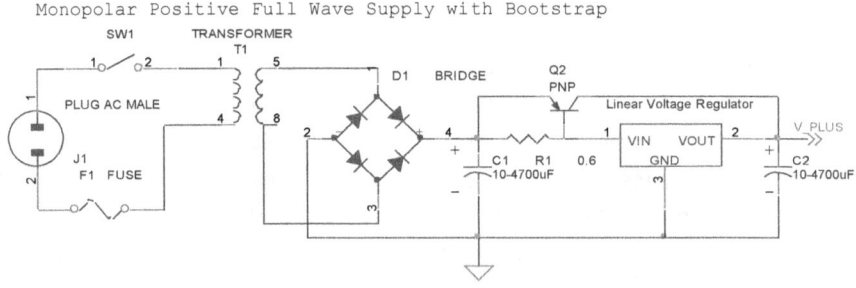

Figure 17: Bootstrapping to increase output current beyond what the linear voltage regulator can normally provide.

Linear Power Supply with +12V, +5V, and -5V outputs

Figure 18 shows a linear regulator that produces fixed voltages at +12 volts, +5 volts, and -5 volts.

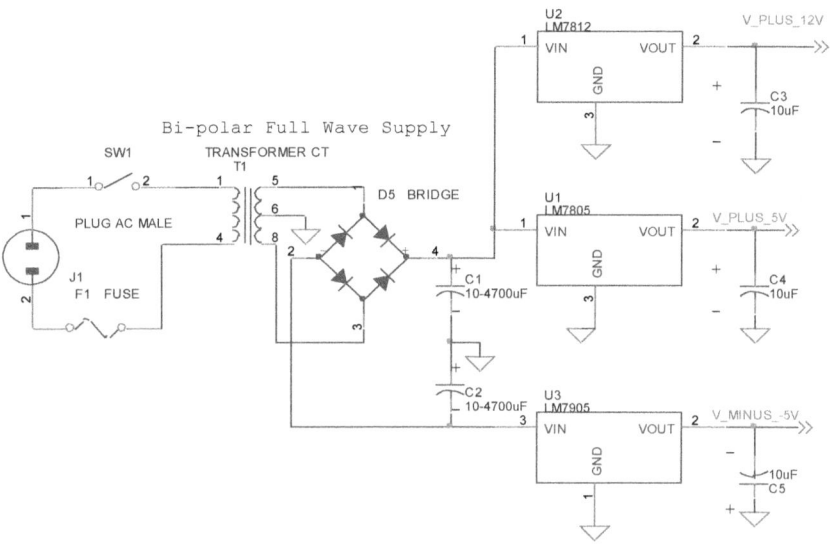

Figure 18: Linear Bridge Power Supply with Two Positive outputs and one negative output.

Switching Regulators

Switching regulators are used for higher efficiencies. All regulators produce heat as a result of the current supplied to the load and the voltage drop across the regulator. Switching regulators produce much less heat because power absorbed in the inductor, is released back into the circuit on each switching cycle. An ideal inductor has no resistance, and therefore consumes no power. In practice, switching regulators are purchased as a module, as shown in Figure 19.

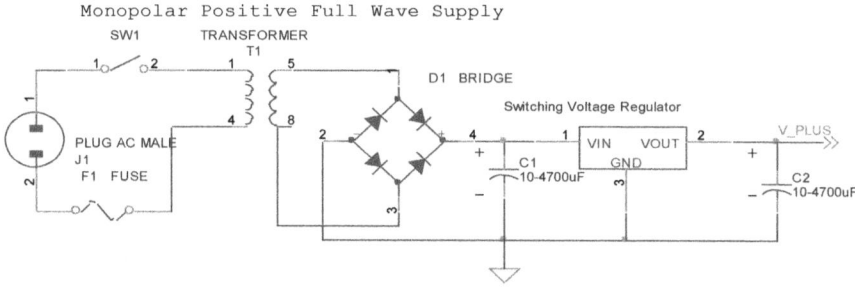

Figure 19: Switching Regulator Model

Boost Switching Regulator

In a boost switching regulator, the output dc voltage is necessarily higher than the input dc voltage. In Figure 21, as SW1 is closed, the current through L1 continues to increase. As SW1 is opened the current continues to flow through D1, to charge the output capacitor. As the current decays, SW1 is closed again to replenish the current through L1. The cycle repeats, often hundreds of thousands of times a second. Switching regulators normally have 10 millivolts to 100 millivolts of ripple at the output, with the frequency of the ripple equal to the switching frequency of SW1.

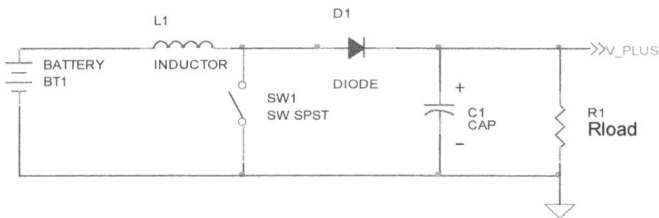

Figure 20: Boost Switching Regulator

Buck Switching Regulator

A buck switching regulator, as shown in Figure 21: Buck Switching Regulator, necessarily has a lower output voltage than the input voltage. As SW1 is closed, current flows through L1 to Rload. When SW1 opens, that magnetic field of L1 starts to collapse in order to maintain the current flow though L1 to Rload. Diode D1 is present so that when SW1 is opened, the current has a closed loop path through L1, C1, and D1. Feedback (no shown) keeps the output voltage at the desired level. Expect 10 millivolts to 100 millivolts of ripple at the switching frequency, at the output (Rload).

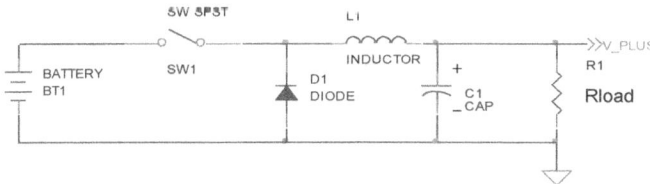

Figure 21: Buck Switching Regulator

Inverting Switching Regulator

Because of the location of the inductor, and the reversal in polarity of D1 and C1, in Figure 22, implements a switching regulator that converts a positive input dc voltage to a negative dc output voltage. As SW1 is closed, it charges L1, which discharges through D1 and C1 when SW1 is opened.

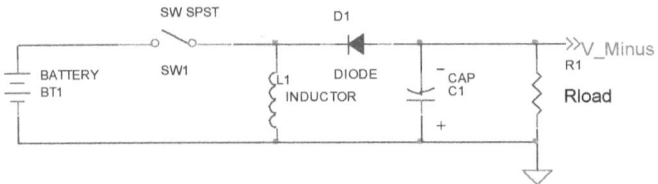

Figure 23: Inverting Switching Regulator

Flyback Switching Regulator

One of the purposes of the flyback switching regulator (see Figure 24: Flyback Switching Regulator) is to provide a very high degree of isolation between the input and the output. This is done by using a transformer, notably called the flyback transformers. Complete isolation means that the dc resistance between the input, V1, and the output V_PLUS is near infinite (1000 mega-ohms or greater). Usually a breakdown voltage of 1500 volts ac root-mean-squared (rms) between T1 primary and T1 secondary is required for the flyback transformer. While the design of the flyback transformer is beyond the scope of this article, the details can be found in **Application Note AN-1024** by International Rectifier.

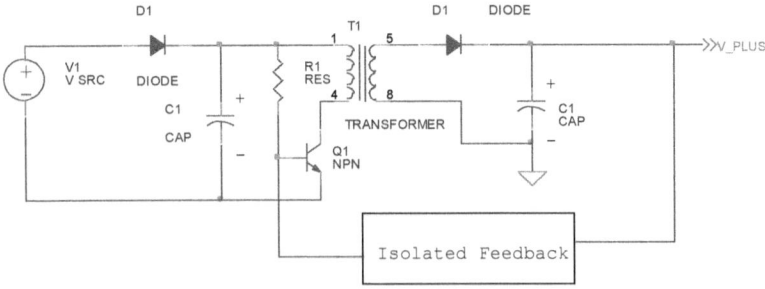

Figure 24: Flyback Switching Regulator

Prototype Buck Switching Regulator with Discrete Components

An elementary switching regulator is shown in Figure 25. There is no adjustable current limiter in this unit, although R1, R2, R3, Q2, R8, R9, C5 and Q4 set the current limit to approximately 10 amps. As you can see, the design is very similar to that of a linear power supply, except that L1, and D1 have been added, and U1 operates in a switching mode as a comparator with a small amount of hysteresis. The switching frequency of this unit varies with the output current drawn by the load. This is an undesirable feature, which is why PWM regulators are used today. With a PWM regulator, the switching frequency is constant and will produce spurs only at known discrete frequencies rather than spurs at all frequencies. The Darlington-connected pass transistor block in the schematic is there twice (in parallel) for robustness. R4 is an internal trim-pot that can set the output voltage anywhere between 5 to 15 volts.

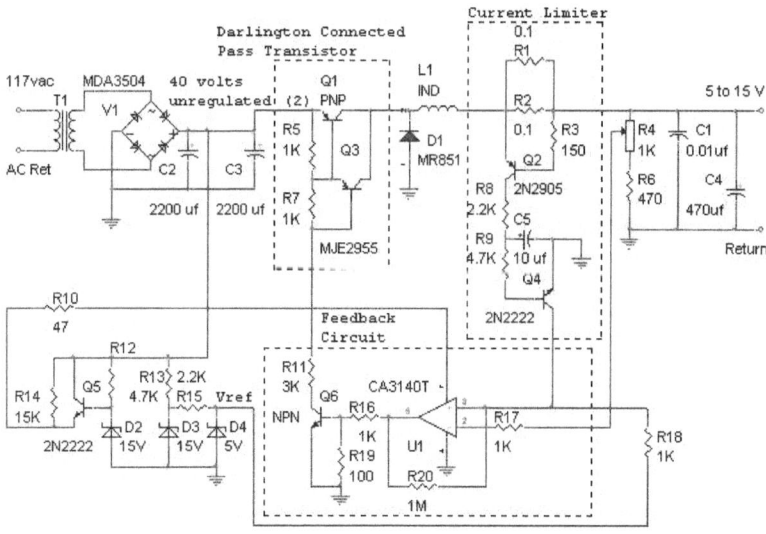

Figure 25: Prototype Buck Switching Regulator

It is always better to use a switched capacitor integrated circuit (Figure 25), rather than try to design one discretely (Figure 26). An integrated circuit solution (Figure 26) will address issues such as constant pulse width modulation (PWM) frequency, soft start, power good (PGOOD) output, under voltage lockout,

17

thermal shutdown, burst mode operation, and overload current limit.

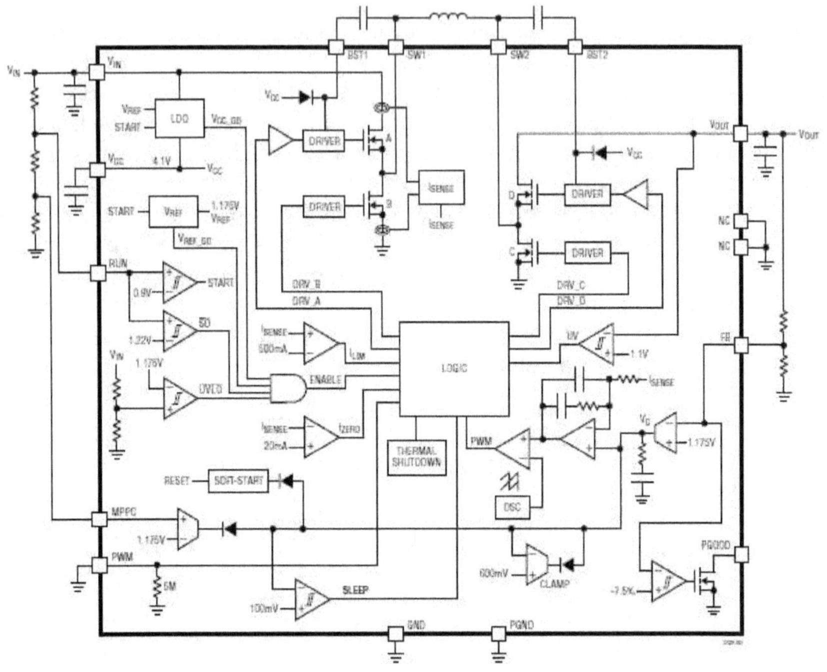

Figure 26: Typical Internal Composition of a Switched Capacitor Integrated Circuit

Emission Requirements of Switching Regulators

Switching regulators that are powered by 110-220VAC, are required to have a common mode line filter as shown in Figure 27: Common mode single-phase line filter to suppress conducted emissions with 110-220 VAC input..

Electromagnetic interference (EMI) is generated by AC/DC and DC/DC switch mode power supplies (SMPS) that generate radio frequency (RF) signals of multiple frequencies. International regulatory bodies have established rules and regulations so that SMPS converter modules comply with such standards are called electromagnetically compatible (EMC) power supplies.

T Federal Communications Commission (FCC), in the United States, is responsible regulates such interference. Any spurious signal greater than 10 kHz in frequency is subject to regulation.

There are two types of EMI emissions. They are radiated and conducted emissions. While radiated emissions are those radiated and coupled through the air, conducted emissions are RF signals coupled through the AC power cable.

The FCC divides electronic equipment using SMPS power supplies into Class A and Class B. Class A is designated industrial environments, and Class B is used in the residential space. Class B is more severe than Class A.

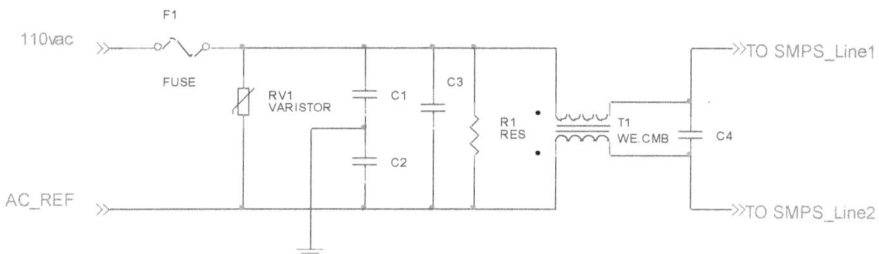

Figure 27: Common mode single-phase line filter to suppress conducted emissions with 110-220 VAC input.

In Europe, CISPR 22 requires certification over the frequency range of 0.15 MHz to 30 MHz for conducted emissions.

The specified limits for the FCC Part 15 are listed in Table 1. Table 2 provides similar limits for CISPR 22. Class B limits are more rigorous than Class A by a factor of 3 (~10 dB).

Table 1: *Field strength limits for conducted and radiated emissions per FCC Part 15 regulations.*

FCC Class A Conducted EMI Limit	
Frequency of Emission (MHz)	Conducted Limit (µV)
0.45 - 1.6	1000
1.6 - 30.0	3000
FCC Class B Conducted EMI Limit	
Frequency of Emission (MHz)	Conducted Limit (µV)
0.455 - 1.6	250
1.6 - 30.0	250
FCC Class B 3-Meter Radiated EMI Limit	
Frequency of Emission (MHz)	Field Strength Limit (µV/m)
30 - 88	100
88 - 216	150
216 - 1000	200
above 1000	200
FCC Class A 30-Meter Radiated EMI Limit	
Frequency of Emission (MHz)	Field Strength Limit (µV/m)
30 - 88	30
88 - 216	50
216 - 1000	70
above 1000	70

Table 2: *Field strength limits for conducted and radiated emissions per CISPR 22 regulations.*

CISPR Class A Conducted EMI Limit		
Frequency of Emission (MHz)	Conducted Limit (dBμV)	
	Quasi-peak	Average
0.15 - 0.50	79	66
0.50 - 30.0	73	60
CISPR Class B Conducted EMI Limit		
Frequency of Emission (MHz)	Conducted Limit (dBμV)	
	Quasi-peak	Average
0.15 - 0.50	66 to 56*	56 to 46*
0.50 - 5.00	56	46
5.00 - 30.0	60	50
CISPR Class A 10-Meter Radiated EMI Limit		
Frequency of Emission (MHz)	Field Strength Limit (dBμV/m)	
30 - 88	39	
88 - 216	43.5	
216 - 960	46.5	
above 960	49.5	
CISPR Class B 3-Meter Radiated EMI Limit		
Frequency of Emission (MHz)	Field Strength Limit (dBμV/m)	
30 - 88	40	
88 - 216	43.5	
216 - 960	46.0	
above 960	54.0	

Decreases with the logarithm of the frequency.

The PM BUS

Many switching regulators today are implementing the PM BUS for managing the switching power supply. The PM Bus is a two wire interface similar to the I2C protocol. By interfacing the switching power supply with a microcontroller, the PM BUS can be used to set the output voltage, measure the input voltage and the output voltage, and even measure the output current. It current applications, it is usually required to meet RoHS II EU "Directive 2011/65/EU, which is a lead free certification.

Figure 28: Switching Converter, 3Vdc –14.4Vdc input; 0.45Vdc to 5.5Vdc output; 3A Output Current with PM BUS management

Example: Application Circuit of a PM BUS Switching Regulator

Requirements:
Vout: 1.8V
Vin: 12V
⊗Vout: 1.5% of Vout (27mV) for worst case load transient
Iout: 2.25A max., worst case load transient is from 1.5A to 2.25A
Vin, ripple 1.5% of Vin (180mV, p-p)

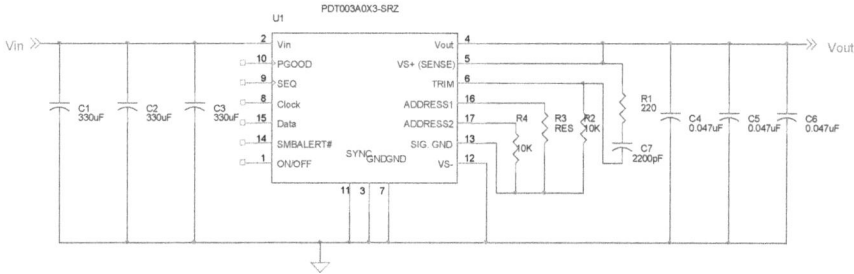

Figure 29: PM BUS Switching Regulator Example. The DATA, CLK and SMBALRT pins do not have any pull-up resistors inside the module. Typically, the SMBus master controller will have the pull-up resistors as well as provide the driving source for these signals.

Example: Application Circuit of a Flyback Regulator

Flyback transformer design is very sophisticated and is usually done with a spreadsheet or other tool specifically used only for this type of design. See **Application Note AN-1024** by International Rectifier for flyback design information.

If you are hired by a large company to design switching power supplies, chances are that you will be tasked with the design of the flyback transformer.

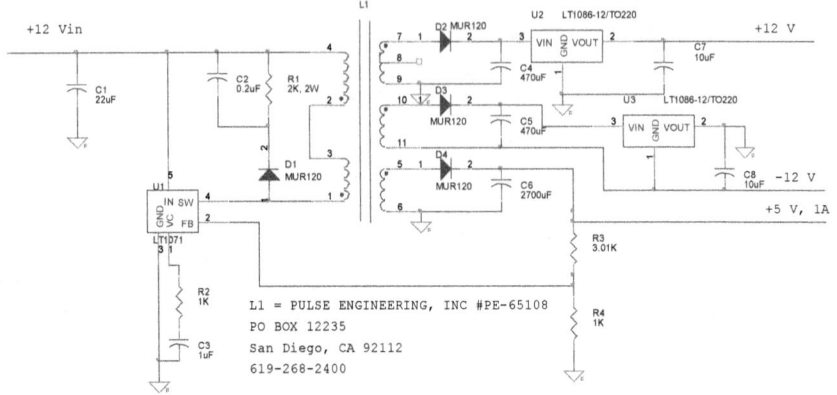

Figure 30: Totally Isolated Flyback Switching Converter (5V to ±15V)

Example: Application Circuit of a Boost Switching Regulator

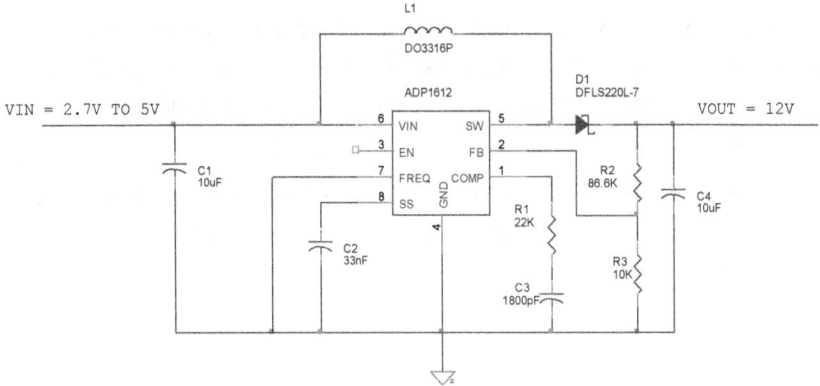

Figure 31: Step-up Boost Switching Converter

Example: The Sepic Switching Converter

A single-ended primary-inductor converter (SEPIC) is essentially a buck-boost converter but has the advantage of having non-inverted output uses a series capacitor to couple energy from the input to the output allowing it to being capable of true shutdown. When the switch is turned off, its output will drop to 0 V.

SEPICs are used in applications where a battery voltage can be below or above that of the regulator's intended output. A single lithium ion battery typically discharges from 4.2 volts down to 3 volts. The SEPIC would be a good choice if other components require 3.3 volts.

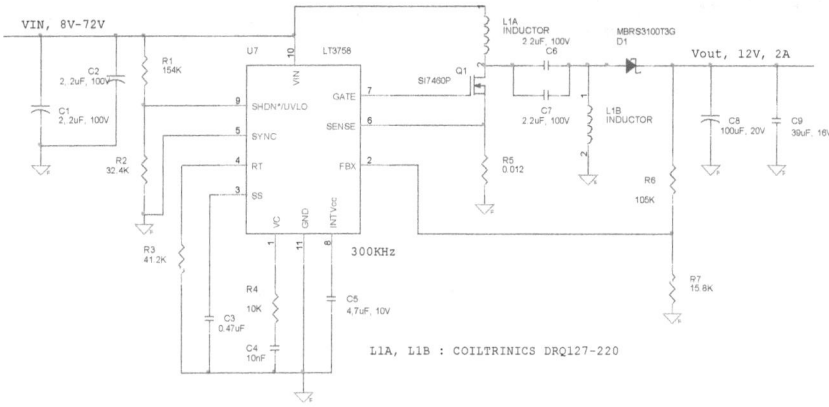

Figure 32: Sepic Switching Power Supply Converter

The LTSpice Switching Power Circuit Simulator

Analog Devices provides a simulation tool that is useful for simulation of switching power supplies. The simulation tool provides a library of switching regulator circuits, so that a reference schematic can be imported, and easily modified.

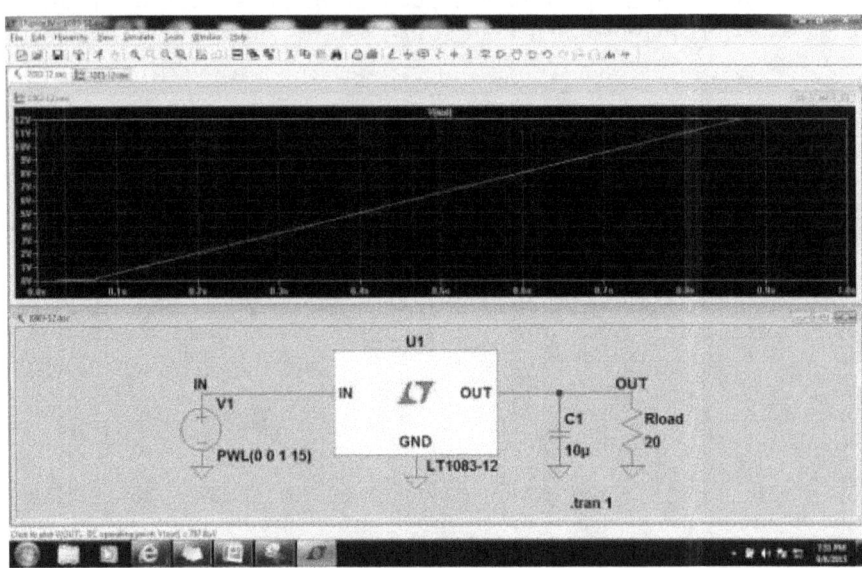

Figure 33: Schematic and Simulation of a Switching Power Supply

The Texas Instruments WEBENCH Design Tool for a Complete Switching Supply Schematic

Texas Instruments provides WEBENCH Designer which is a powerful software algorithm with visual interfaces that deliver complete applications in seconds. Simply specify the input voltage range, and the output voltage with maximum current and the design is done in seconds.

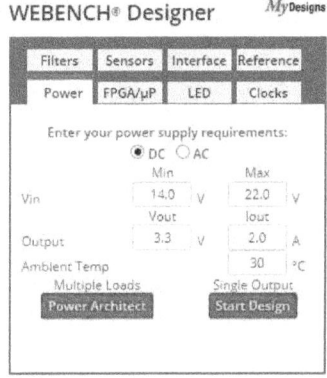

Figure 34: The Texas Instruments WEBENCH design interface for switching power supplies.

The ADIsimPower Tool from Analog Devices

ADIsimPower will generate schematics, provide a bill of material (BOM), and includes thermal analysis.

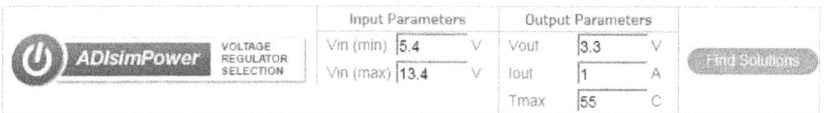

Figure 35: ADIsimPower user interface

Special Circuits

Passive Voltage Doubler

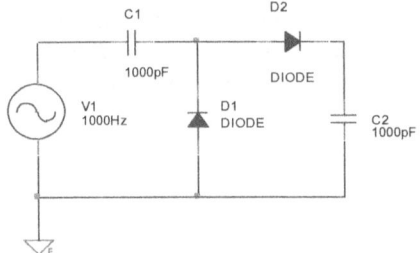

Figure 36: Passive Voltage Doubler with AC Source

Passive Voltage x8

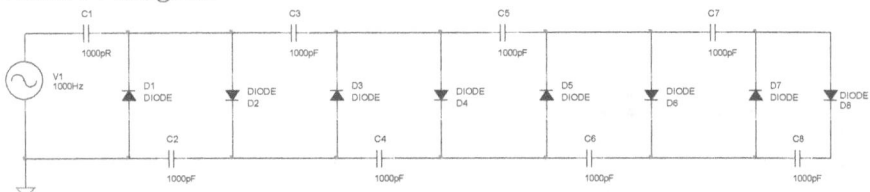

Figure 37: Passive Cockcroft-Walton Voltage x8 Circuit with AC Source

NTSC televisions have a cathode ray tube that requires a high voltage. This high voltage is generated by a flyback circuit in conjunction with a passive voltage tripler, similar to the figure above. One such tripler is ERO BG 2032-642-3003M.

Figure 38: Commercial Voltage Tripler designed to produce the acceleration voltage, focus voltage and second-grid voltage in NTSC color television receivers

A 12VDC to 110VAC Converter for Automobiles

An inverter for an automobile usually plugs into the cigarette lighter and converts 12 volts dc to 110 volts ac. The conversion was done at 60Hz and required a very large and heavy transformer for the final conversion. A typical schematic of the old style inverter is shown below.

A much smaller transformer can handle the same power conversion if a higher frequency is used. For a given wattage, the ferrite core high frequency transformer operates with lower flux densities than the larger iron core transformer.

In order to use the smaller transformer, the input dc must be chopped at 50KHz, then the smaller transformer is used to step up the voltage at 50KHz. This voltage is then rectified and the resulting high dc voltage is pulse width modulated to simulate a 60Hz sine wave.

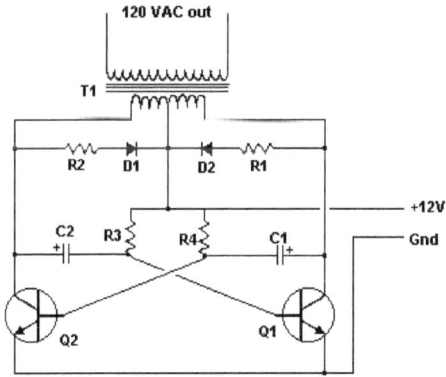

Figure 39: Old Style Automotive 12VDC to 110VAC Inverter.

Flux density (Bmax), a key design factor of a transformer, is a function of both cross sectional area of the core and the frequency.

Bmax = Vrms × 108/4.44N × Ac × F for sine waves

Bmax = Vpeak × 108/4N × Ac × F for square waves

In these equations: V - voltage (volts), N - winding's turns, Ac - core's cross-sectional area (sq.cm),
F- frequency (hertz). Desired flux density is achieved by reducing the cross sectional area
of the core but increasing the frequency. The figure below shows the size reduction that can be achieved by using a higher switching frequency.

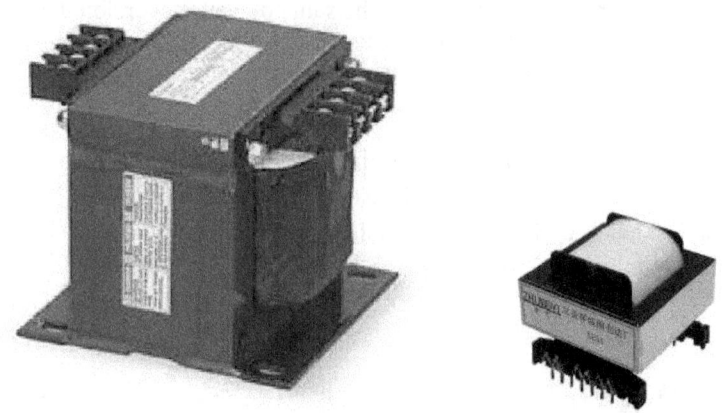

Figure 40: The smaller transformer handles the same power at 50KHZ as does the larger transformer at 60HZ.

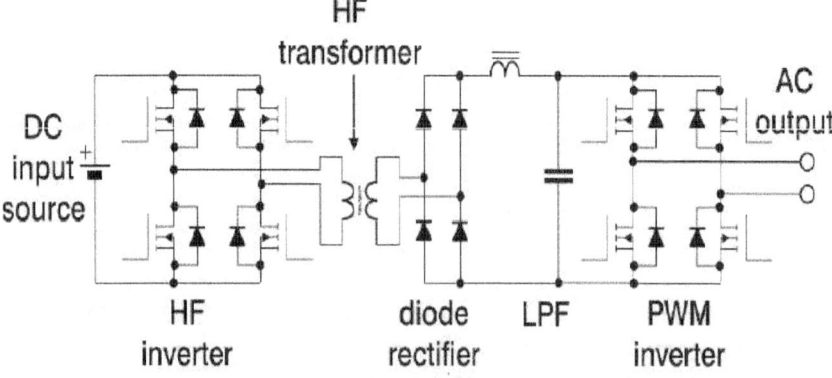

Figure 41: Lightweight automobile 12VDC to 110VAC inverter block diagram

$V_{spwm.1}$ NV_B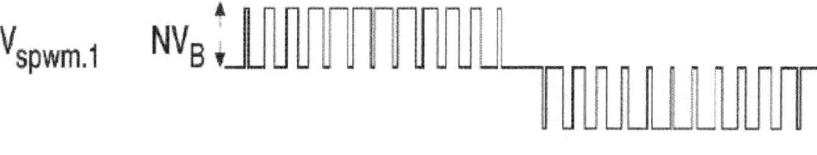

V_{spwm} $\sqrt{2}220V$

Figure 42: Pulse Width Modulation at 50KHZ produces a simulated sine wave at 60HZ.

Brokaw bandgap reference

In order to operate at a stable voltage, a switching, or linear voltage regulator must have a stable reference voltage. The Brokaw bandgap reference is a voltage reference circuit commonly used in integrated circuits, having an output voltage around 1.25 V with very little temperature dependence.

The circuit uses an internal voltage source that has a positive temperature coefficient and an opposing internal voltage source that has a negative temperature coefficient. By summing these two together, the temperature dependence can be canceled. In addition, either of these two internal sources can be used to sense temperature.

In the Brokaw bandgap reference, the circuit uses op-amp negative feedback to force a constant current through the two bipolar transistors that have different emitter areas. By the Ebers–Moll model of a transistor,

- The transistor with the larger emitter area has a smaller base–emitter voltage for the same current.
- The base–emitter voltage for each of the transistors has a negative temperature coefficient that decreases with temperature.
- The difference between the two base–emitter voltages results in a positive temperature coefficient that increases with temperature.

The output is the sum of the base–emitter voltage difference with one of the base–emitter voltages. With proper component choices, the two opposite temperature coefficients will cancel each other exactly and the output will have no temperature dependence, providing a stable reference voltage.

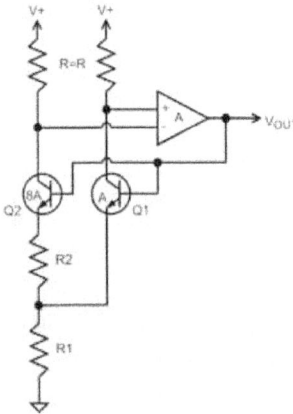

Figure 43: Brokaw Bandgap Reference with Vout = 1.25 volts

Capacitive Linear Power Supply

A **capacitive power supply** is a type of power supply that uses the capacitive reactance of a capacitor to reduce the mains voltage to a lower voltage

Given the circuit topology below with the input voltage consisting of a square wave and an output load of 5 volts at 10 ma, the problem is to find the component values for the resistors and capacitors. Assume that the input voltage can range from ±40 to ±120 volts with a wave shape as shown below.

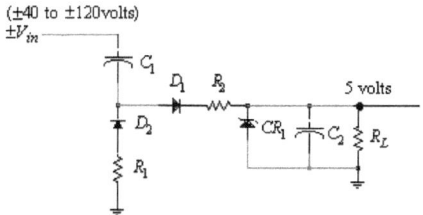

Figure 44: Low current power supply circuit

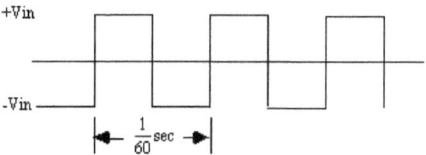

Figure 45: Square wave input voltage, Vin. Range will be ±40 to ±120 volts.

To derive the design equations, it will be necessary to assume that C1 fully charges on each half cycle. This assumption will be validated later.

Design Derivation

Since an output voltage of 5 volts is specified, the required zener (CR1) voltage is also 5 volts.

The figure below shows the model to compute the voltage drop across C1 and the end of the negative half cycle. Similarly, the figure below shows the model for computing the voltage drop across C1 and the end of the positive half cycle.

Using the relation,

$$Q = C_1 V$$

the total charge transferred over one full cycle is

$$C_1(2V_{in}-6.2)$$

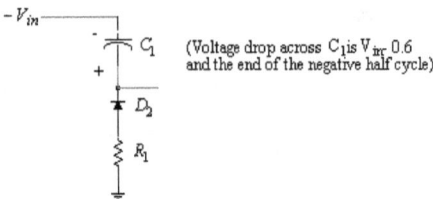

Figure 46: Voltage drop across C1 at end of negative half cycle.

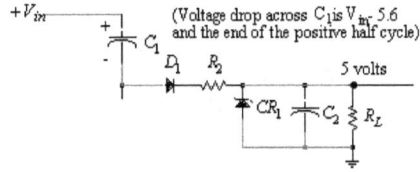

Figure 47: Voltage drop across C1 at end of positive half cycle.

But current is charge per second, so the average current flow through D1 over one second is

$$I = 60C_1(2V_{in} - 6.2) \text{ amps}$$

since there are 60 cycles per second.

The desired maximum output current is 10 milliamps. Vin can range from ±40 to ±120 volts (square wave). Using the worst case input voltage (±40 volts), C1 can be computed from equation (3) as

$$C_1 = \frac{0.01}{(80 - 6.2) * 60} = 2.26 \text{ microfarads}$$

C2 should be large enough to hold the output voltage to within 5% of 5 volts over the negative half cycle using the maximum load of RL = 500 ohms (for 10 milliamps). Therefore

$$0.95 * 5 = 5e^{-\frac{1}{R_L C_2} t}$$

or

$$C_2 = \cfrac{1}{-\ln\left(\cfrac{4.75}{5}\right)*500*120} = 325 \text{ microfarads}$$

using 1/120 second for the time duration of the negative half cycle.

To determine R2, assume the maximum peak instantaneous power that the zener diode can handle is 10 watts. Also assume that the average power that the zener can handle is 0.2 watts. (The zener diode will have to be selected to meet these specifications.) Since the output current is already determined to be 10 milliamps, maximum, the average power to the zener with an input voltage square wave of ±40 volts is

0.01 amps *5 volts = 0.05

but for the maximum input voltage square wave of ±120 volts, the average power to the zener will be approximately

0.03 amps * 5 volts = 0.15 watts

However the peak surge current to the zener is approximately given by

$$\frac{2V_{in}}{R_2} = I_{max}(peak)$$

since the input voltage seen by R2 is Vin plus the voltage across C1, or approximately 2 Vin total. From equation (9), for Vin = 120 volts (worst case), and an assumed surge current rating of 2 amps on the zener, the minimum value of R2 becomes 120 ohms.

There is another limiting factor on R2. From our initial assumption C1 must be fully charged in each half cycle of the input voltage, Vin. In five time constants, C1 will charge to 99.6% of its final voltage. Therefore from the time constant R2-C1, R2 may be determined from

$$R_2 \leq \frac{\frac{1}{120} \frac{1}{5}}{2.26 * 10^{-6}} = 737\Omega$$

The bounds on R2 become

$$120 \leq R_2 \leq 737$$

R2 can be selected near the high end to limit the surge current to the zener, therefore we choose R2 = 680 ohms, limiting the zener surge current to 0.35 amps. For equal time constants on both the positive and negative half cycles in the input voltage, choose

R1 = R2 = 680 ohms.

The maximum power dissipated in R1 and R2 occurs when Vin = ±120 volts. Then the average current is approximately 3 times higher than when Vin was ±40 volts. The average current with Vin = ±120 volts is approximately 0.03 amps. The maximum dissipated power in R1 and R2 is

$$(0.03)^2 680 = 0.612 \text{ watts}$$

The final design for a 5 volt, 10 milliamp supply is given in Figure 5. Although it is not shown, a RF choke should be added in series with C1 as an RF block. The spice simulation plot is shown below.

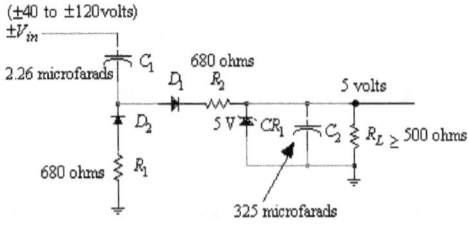

Figure 48: Final design of a Capacitive Power Supply

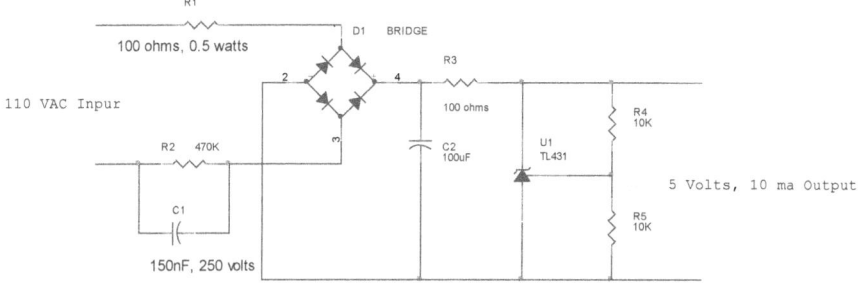

Figure 49: An alternative implementation given without the design equations.

Spice Simulation File

```
EXAMPLE - Capacitive coupled supply
.WIDTH OUT=80
.OPT ITL5=10000 ACCT NOECHO NOPAGE OPTS RELTOL=.001
.TRANS .5S 1S .001S
VPULSE 1 0 PULSE (-40V 40V 100NS 100NS 100NS .00833S .016666S)
C1 1 2 .00000225
D1 2 4 DIODE
D2 3 2 DIODE
R1 3 0 680
R2 4 5 680
R3 5 0 500
C2 5 0 .000325
D3 0 5 ZENER
.MODEL DIODE D(RS=1 VJ=0.6)
.MODEL ZENER D(RS=1 VJ=0.6 BV=5 IBV=1)
.PROBE
.END
```

Based on the preceding example, the generalized design equations for a 60 Hz square wave ac input can be written as follows.

Input parameters:

Minimum ac square wave input voltage = Vin(min)

Maximum ac square wave input voltage = Vin(max)

Desired dc output voltage = Vout

Maximum desired dc output current = Iout

Component values:

Zener (CR1) voltage = Vout

$$C_1 = \frac{I_{out}}{\left[2 V_{in(min)} - \left(V_{out} + 0.6 + 0.6\right)\right] * 60} \text{ farads}$$

$$C_2 = \frac{1}{\left[-\ln(0.95)\right] * (2*60) * \left(\frac{V_{out}}{I_{out}}\right)} \text{ farads}$$

Maximum possible value of R_1 and R_2 is $R_1 = R_2 \leq \dfrac{(2*60)^{\frac{1}{5}}}{C_1}$ ohms

Average power rating required for zener $\geq \left[\dfrac{V_{in(max)}}{V_{in(min)}} * I_{out} * V_{out}\right]$ watts

Peak power rating required for zener $\geq \left(\dfrac{2 * V_{in(max)}}{R2} V_{out}\right)$

Power rating of R_1 and R_2 = $\left(\dfrac{V_{in(max)}}{V_{in(min)}} I_{out}\right)^2 R_1$

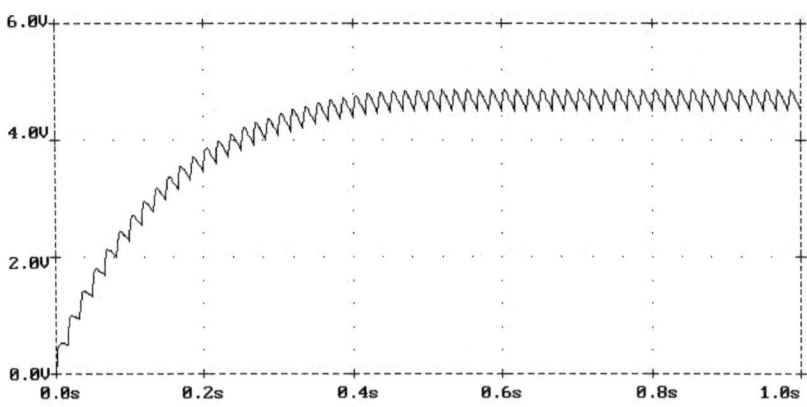

Figure 50: Spice Simulation of Capacitor Power Supply

Due to the absence of electrical isolation, the Capacitive Linear Power Supply must be encapsulated and isolated to avoid direct (galvanic) contact with the users.

Switched Capacitor Converters

A switched capacitor inverter is a charge-pump voltage inverter which can generate a negative supply from a positive input voltage. This is done WITHOUT the need for an inductor, either internal to the integrated circuit, or external. Generally only two small external capacitors are needed for the charge pump. Outputs with greater than a 99% efficiency are achievable. The switched capacitor inverter is limited to small output currents. The application shown below can only provide up to 25ma.

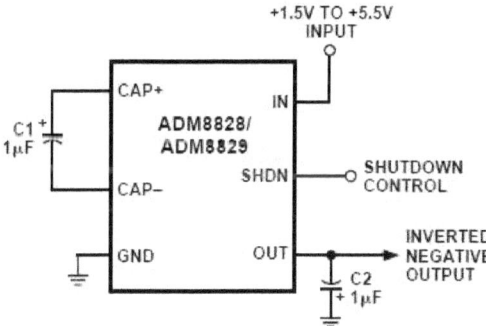

Figure 51: Typical Charge Pump Application used to Convert a Positive Voltage to a Negative Voltage.

A Cathode Ray Tube Power Supply Example

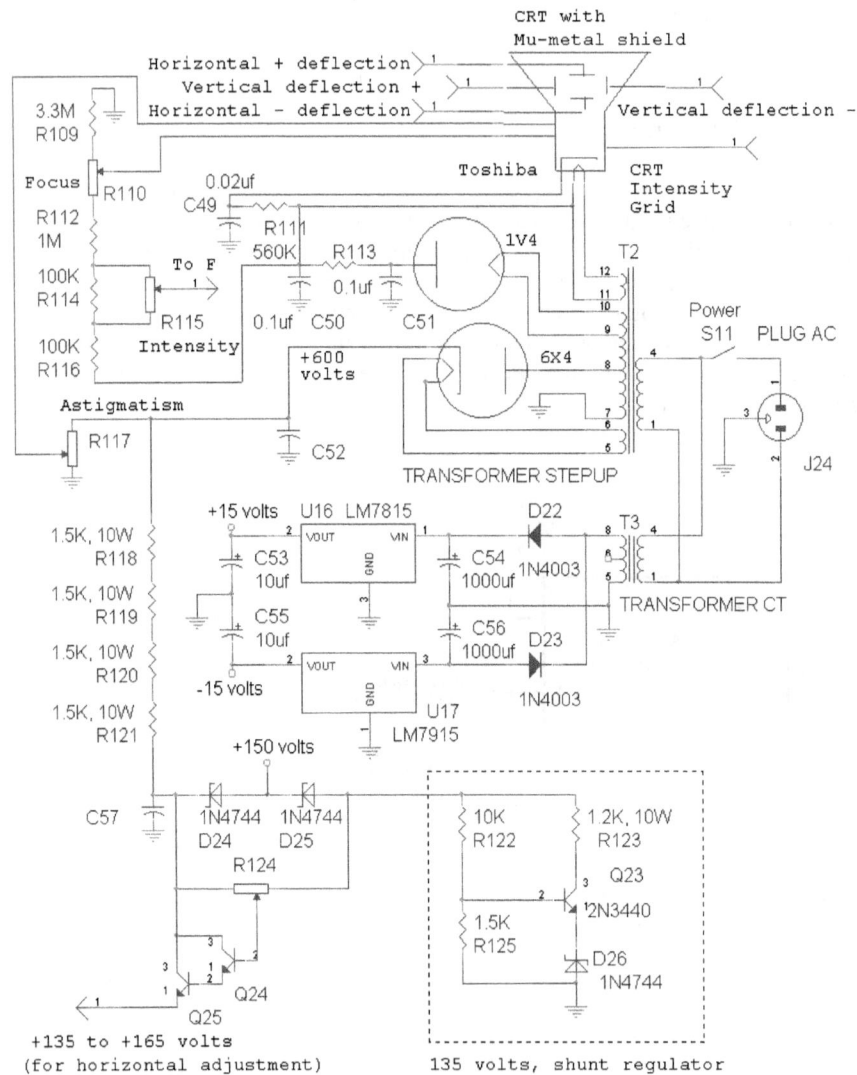

Figure 52: Example Circuit: Cathode Ray Tube High and Low Voltage Power

Grounding

There's only one true ground. That is the conductive mass of the earth, whose electric potential at any point is traditionally taken as equal to zero volts."

The other grounds are local:

- Frame Ground (FG)
- Chassis Ground (CG)
- Logic, Analog, Digital ground (LG)
- Battery Return (BTRN)

Radio receivers, biomedical equipment, telecommunication equipment, etc., must be able to detect very small voltage differences to operate properly. The signal voltage is many magnitudes less than the power voltages used in the system. To improve sensitivity the frame ground, chassis ground, and logic (electronic circuit) ground are isolated from each other, but connect directly to earth ground.

"Generic Telecommunications Bonding and Grounding for Customer Premises", ANSI/TIA-607-B, is by far the most comprehensive telecommunications standard explaining bonding and grounding.

Overall Frame Bonding

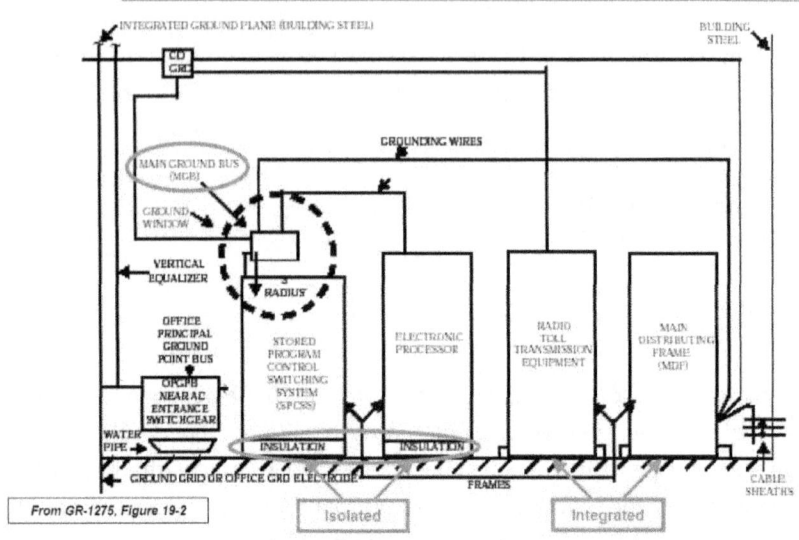

Figure 53: Chassis Grounding Practice from Telcordia GR-1275

HiPot Isolation Safety Testing

High-potential (HiPot) safety testing is required before a product can be commercialized. Standard IEC60950-1 applies to ac-dc power supplies and requires power supplies to pass an input-to-output isolation of 3 kV ac. Standard IEC60601-1 applies to medical power supplies and testing is done at 4KC ac.

These are stress tests for the main transformer only. It requires removal of the main transformer from the circuit to test the isolation between the primary and the secondary of the transformer.

Hi-Pot testing is also required between a product's electronic circuit and its isolated metallic shielding.

Power Generation for Small Scale Systems

Solar panels generate power from sunlight by converting sunlight to electricity with no moving parts, zero emissions, and no maintenance. Generally, solar power is far more expensive than using power from the local utility power company. The solar panel is a set of individual silicon cells that generate electricity from sunlight. Sunlight produces an electrical current when it strikes the surface of the thin silicon wafers. A single solar cell produces only about 1/2 of a volt. A typical 12 volt panel about 25 inches by 54 inches (1350 square inches) will contain 36 cells wired in series to produce about 17 volts peak output. Two 12 volt groups of cells 36 each can be wired in series allowing the solar panel to produce 24 volts. When under load (charging batteries for example), this voltage falls to 12 to 14 volts (for a 12 volt configuration) for 75 to 100 watts for a panel of 2700 square inches.

Solar panels can be (1) Monocrystalline, (2) Polycrystalline, or (3) Amorphous.

The components of a solar power system include

a) Storage batteries are needed to store energy, otherwise you would only have power when the sun was shining.
b) A charge controller is needed to protect the batteries from over charging
c) The power inverter converts the storage battery power into the 110 volts AC that runs your appliances.

Figure 54: House Boat with Solar Power

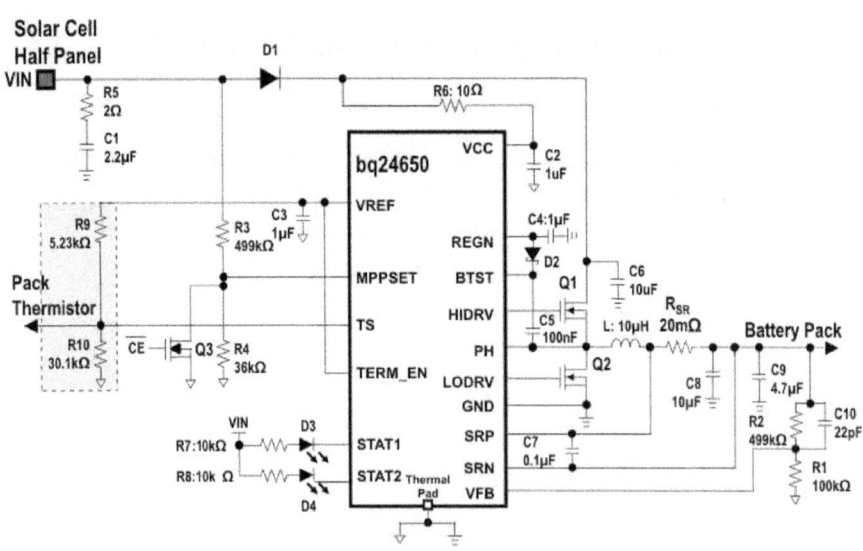

Figure 55: Charge Controller for a Solar Power System

Wind Turbine

Most wind turbines will produce 12 VDC. Therefore the additional components include

a) Storage batteries are needed to store energy, otherwise you would only have power when the wind was blowing.
b) A charge controller is needed to protect the batteries from over charging
c) The power inverter converts the storage battery power into the 110 volts AC that runs your appliances.

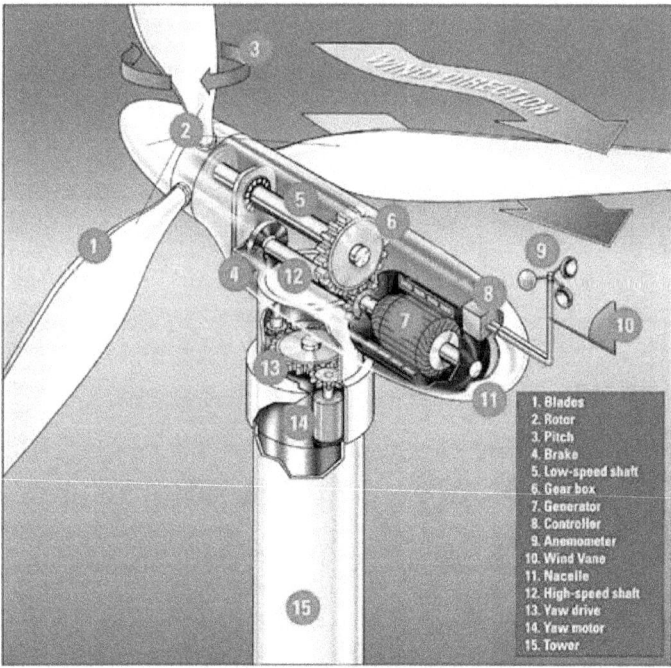

1. Blades
2. Rotor
3. Pitch
4. Brake
5. Low-speed shaft
6. Gear box
7. Generator
8. Controller
9. Anemometer
10. Wind Vane
11. Nacelle
12. High-speed shaft
13. Yaw drive
14. Yaw motor
15. Tower

Figure 56: Wind Turbine working Mechanism

Hydroelectric

A growing do-it-yourself (DIY)- communities have endeavored to build hydroelectric plants from old water mills, steams, using kits or from scratch.

A pico-hydro generator made can be built from common PVC pipe and a modified Toyota alternator in a large bucket. The generator can provide power to communities without access to the electricity grid in developing countries. The pico-generator is designed to be made by anyone with basic skills and can be built inexpensively. The Toyota alternator used in the generator is converted to a permanent magnet alternator allowing it to generate power at low RPMs.

a) Storage batteries are needed to store energy, otherwise you would only have power when the water was flowing.
b) A charge controller is required to protect the batteries from over charging
c) The power inverter converts the storage battery power into the 110 volts AC that runs your appliances.

Figure 57: Hydroelectric Do-It-Yourself System

Thermoelectric Generation

A thermoelectric generator (also called Seebeck generator) is a device that convert heat directly into electrical energy, using a bimetallic junction.

Thomas Johann Seebeck, in 1821, found that a temperature difference formed between two dissimilar conductors generates a voltage. In 1834, Jean Charles Athanase Peltier found the reverse effect. Running an electric current through the junction of two different metals, depending on the direction of the current, could produce a heater, or a cooler.

Thermoelectric generators have no moving parts. Space probes, including the Mars *Curiosity* rover, usually generate electricity using a radioisotope element that produces heat. The heat difference, between the radioisotope and the surrounding environment, is used by the thermocouple to produce electricity.

Typical efficiencies are around 5–8%.

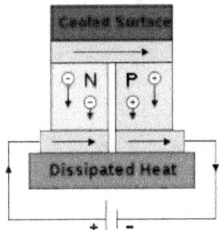

Figure 58: Thermoelectric heater/cooler with applied voltage

A thermopile is a bunch of thermocouples connected in series, which allows them to generate more voltage than a single thermocouple ever could.

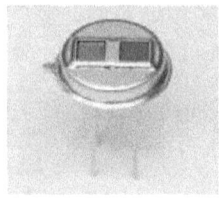

Figure 59: T11722-01 Thermopile

The T11722-01 is a dual-element thermopile detector in a metal TO-5 package. It provides highly accurate CO_2 detection.

The Piezoelectric Effect

Piezoelectric crystals generate an electric voltage in response to applied mechanical stress, or vibration. Piezoelectric crystals can act as the ignition source for cigarette lighters and have been used in record players for decades. The grooves of the record move the stylus which bends the crystal to generate an electrical signal.

Figure 60: The stylus (red) transfers vibrations, from the mechanical variations in the records grove, to the piezoelectric cartridge (gray), which generates audio electrical signals

Nanogenerators

A nanogenerator consists of vertically grown ZnO nanowire, producing electric energy from three known sources including friction, shaking and vibrating (piezoelectric), and temperature gradients (thermoelectric/pyroelectric).

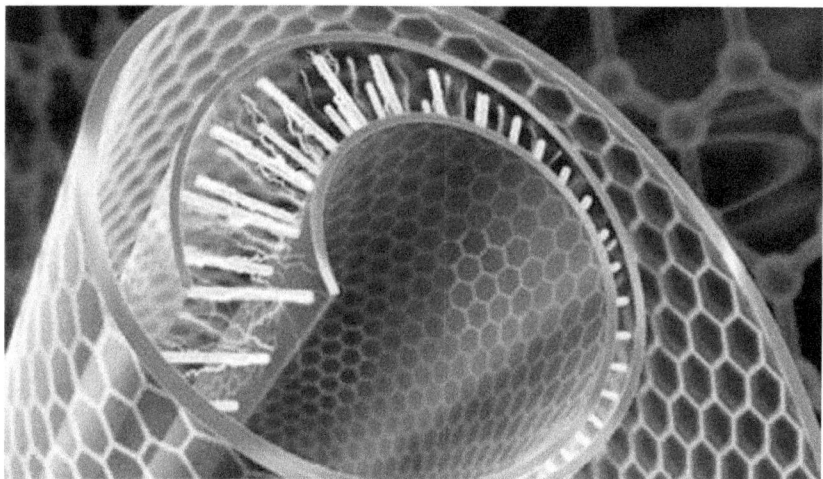

Figure 61: Georgia Tech Nanogenerator, 2006

Battery

A battery converts chemical energy into electrical energy. There are primary batteries, which are not rechargeable, and secondary batteries which are rechargeable. A third type of battery is a criminal offense involving unlawful physical contact. Batteries were known to be used 2000 years ago in the Roman empire, and were possibly used for electroplating.

The following table list several ultra high density batteries. For reference, 454 grams is equal to one pound.

There is no such thing as an ac battery.

Table 3: Ultra High Density Batteries

No	capacity mAh	voltage	Internal mΩ resistance	Weight gr.	Thickness	Width	Length
ABLP5274J0HG	8900	3.7	8	140	5.2	70	170
ABLP7374J0HG	12600	3.7	4.5	196	7.3	70	170
ABLP9255275HG	21000	3.7	3.5	313	9.2	55	271

No	capacity mAh	voltage	Internal mΩ resistance	Weight gr.	Thickness	Width	Length
ABLP6059190HG	10300	3.7	10	159	6.0	59	190
ABLP1158150HG	13500	3.7	3	207	10.5	58	150

No	capacity mAh	voltage	Internal mΩ resistance	Weight gr.	Thickness	Width	Length
ABLP8474J0HG	14400	3.7	8	222	8.4	70	170
ABLP12058150HG	15500	3.7	3	236	11.5	58	150

No	capacity mAh	voltage	Internal mΩ resistance	Weight gr.	Thickness	Width	Length
ABLPA655275HG	23880	3.7	3.5	364	10.6	55	271

High density batteries must be charged under very controlled conditions and it is recommended to use an integrated circuit specifically designed to charge a specific type of battery.

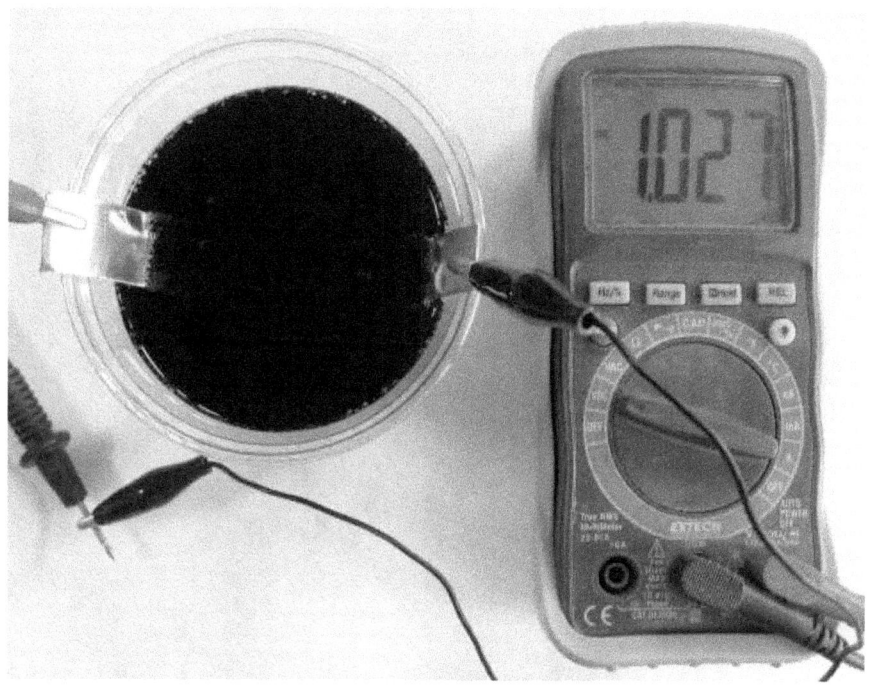

Figure 62: Coca-Cola Battery, made with copper and zinc strips, produces 1.0 volts

Gasoline Powered Electric Generator

Figure 63: Gasoline Powered Electric Generator

This section would not be complete without including gasoline powered generators. Such generators are easily purchased for as low as $189.99 from places like Home Depot. These systems usually include a 12VDC output, and also multiple outlets for 110VAC.

Most cheap portable generators have alternators with fixed excitation. When this type of alternator is loaded, its output voltage, V_{out}, drops due to its internal impedance. This impedance comes from leakage reactance, armature reactance and armature resistance. More expensive models use an automatic voltage regulator (AVR) to maintain V_{out} within tight limits. The AVR controls the output by sensing the voltage at the power-generating coil and comparing this voltage to a stable reference. The error signal then adjusts the average value of the field current.

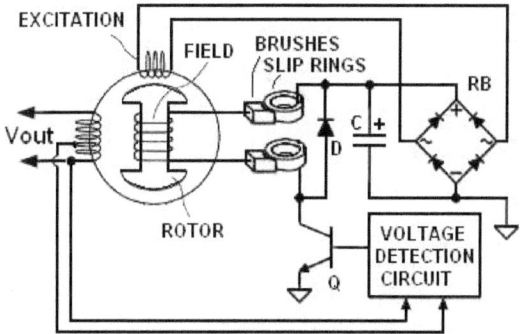

Figure 64: Generator Block Diagram

The block diagram shows the basic concepts to stabilize the output with self-excited alternators. When the rotor is rotated by the gas powered engine, an AC voltage is generated in the excitation winding. This AC voltage is converted a DC voltage by the rectifier diode bridge and capacitor C. The detection circuit compares a voltage representing V_{out} with a reference voltage and turns ON and OFF the transistor Q. When Q is turned ON, a current flows through the field winding. When Q is turned OFF, the field current is decaying while continue flowing via free-wheeling diode D. The rotor sometimes is made from a small permanent magnet to provide detectable baseline current when Q is turned OFF. V_{out} can be regulated, by properly varying duty cycle of the operation of the transistor Q. The transistor, Q, can also operate in linear mode, but its heat dissipation will increase.

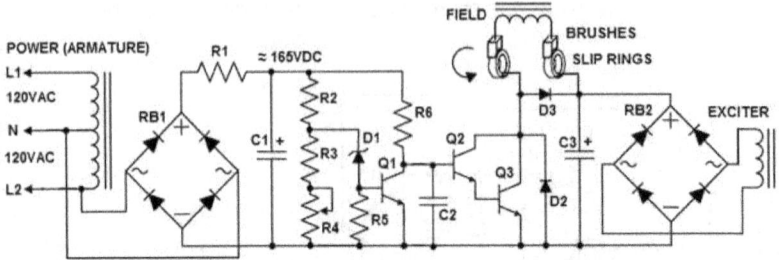

Figure 65: Generic AVR Voltage Regulator

The rectifier, RB1, with capacitor C1 produces DC level close to the peak of V_{out}. A small resistor R1 limits C1 charge current. This resistor may be omitted and replaced with a short. If the divider R2-R3-R4 is properly set, when V_{out} is below its required value, Q1 will be turned OFF and Q2 will be forward biased via R6, and Darlington pair Q2, Q3 will energize the field winding. When V_{out} rises and voltage at the cathode of D1 exceeds approximately Vz+0.7 volt, Q1 opens and shuts down both Q2 and Q3.